"Bankole has clearly built on the foundation laid by my brother Alex in Roots"
George W.B. Haley, Former US Ambassador to The Gambia

A Matter of Black Transformation

Bankole Thompson

Foreword by Dr. Bernard LaFayette

Director, Center for Nonviolence and Peace Studies, University of Rhode Island, who was appointed National Program Administrator of the Southern Christian Leadership Conference, and National Coordinator of the 1968 Poor Peoples' Campaign by Dr. Martin Luther King, Jr.

A Matter of Black Transformation

Bankole Thompson

© Bankole Thompson, 2007

All rights reserved.

ISBN: 0-9760755-7-1
978-0-9760755-7-8

Library of Congress Control Number: 2007930086

No part of this book may be reproduced, stored in a retrieval system or transmitted in any form or by any means - electronic, mechanical, digital, photocopy, recording – except for brief quotations for teaching, discussions, reviews and articles, without the written permission of Bankole Thompson.

Edited by: Paschal Eze

Cover photograph by: Geronimo Patton

Published by:

Read4Life **Books**
An Imprint of Christian Book Outreach, Inc.
P.O. Box 5145 Coralville Iowa 52241 USA
Phone/Fax: 1-319-351-9695

Table of Contents

Dedication 6
Acknowledgement 8
Foreword 11
Preface 13
Chapter One 18
    Pushing frontiers of Black education 18
    Creating and preserving our institutions 21
    Teaching Black children entrepreneurship 25
    Let's have African-centered education 27
    Black education in the Internet age 34
Chapter Two 37
    Help! Black economic power in chains 37
    The Black American challenge 42
Chapter Three 46
    Some Lessons from China 46
Chapter Four 53
    Whither African Growth and Opportunity Act (AGOA)? 53
Chapter Five 60
    The human cost of American chocolates 60
Chapter Six 68
    Enough of the natural resource-driven wars 68
    Black genocide 75
After-word 79
Appendix ii 92
    Interview with Adrianne George, Black American entrepreneur in Europe 92
Appendix iii 98
Interview with Tony Mottley, producer American Black Journal, PBS 98
Appendix iv 104

Interview with US Representative, Carolyn Cheeks Kilpatrick (D-Detroit), head of the Congressional Black Caucus............................ 104
Appendix v.. 110
Interview with Kenneth Harris, founder International Detroit Black Expo................. 110
Appendix vi... 117
Interview with Howard Starks, lecturer, Wayne State University Department of Africana Studies 117
Bibliography 121
About the author 123

Dedication

This book is dedicated to the late Hannah Thompson, "Mommy" my grandmother who raised me and instilled the spirit of advocacy and action in me. Mommy, because of your consciousness, wisdom and love, I am able to see the world from a different lens. Your wisdom allows me to use discretion without compromising my integrity. I am what I am today because of you, Mommy. Thank You!

Acknowledgement

This work could not have come to fruition without the earnest assistance of a few people who believed in me and respect my position on the state of the Black World.

First is Ambassador George Haley who thought this work was important to come out at the 30th anniversary of *Roots*, the 1977 masterpiece of his brother, Alex Haley. Ambassador Haley who has been supportive of my efforts believes this book makes a strong case for sustained socio-economic ties between Black Africa and Black America.

Second is Shanita Michelle of **Read4Life Books Iowa**, my publishers, who pressed me to complete this work in time for the *Roots* anniversary. Even though she did not give me a breathing space, I found her insistence on finishing this work to be very helpful. I'm glad I was able to finish the project on time.

My book editor, Paschal Eze, maintained a sharp editor's eye on this work, following my thoughts from chapter to chapter. I could not have finished this book without his remarkable book doctor's work and very useful comments and suggestions.

I'm thankful to Sam Logan, publisher of the Michigan Chronicle, and his chief operating officer,

Karen Love, for the confidence they've placed in my journalistic work and editorial judgment at the Chronicle. Such confidence has enabled me - with the strong backing of the editorial team - to carry out public/community journalism that places the people's interest in the forefront, and lives the true meaning of what a Black newspaper is about. Working with editorial team members - Marcus Amick, Terry Cabell, Melody Moore, Steve Holsey, Cornelius Fortune, Patrick Keating - and the production staff of the paper has been a wonderful experience. These sons and daughters of the ink make things happen at the Chronicle.

My good friend, Dr. Bernard LaFayette of the University of Rhode Island and a prominent deputy of the civil rights movement has been supportive. He is passionate about Black transformation and is involved in teaching the non-violence philosophy of Dr. Martin Luther King Jr. around the world - including South Africa and Nigeria. When I reached out to him to do the foreword for the book, he gladly accepted my request, despite his busy travel schedule as an international academic.

I thank all those who read my weekly reports and columns in the Michigan Chronicle. You've kept me on my toes, bringing news and opinions to you from a different perspective.

And finally I want to thank Alex Haley for his masterpiece, *Roots*, in tracing the historical journey of Blacks in America. My relationship with the

Haley family, which began many years ago, has culminated into this book. *Roots*, according to Alex, was the saga of an American family. Thirty years after *Roots*, the problems we face as Black people around the world still speak to the saga of the Black race.

Foreword

The issue of Black empowerment has been at the forefront of challenges for America for the last three centuries. Blacks as well as Whites have shared a burden of struggling to find the keys to unlock the grip of strangling poverty, particularly affecting the Black community.

Questions related to the motive and intent of slavery, segregation and the aftermath have plagued us. There have always been fragmented theories about why certain conditions exist and conspiratorial motives that restrict the upward mobility of Blacks in America.

However, this book, for the first time, lays out in clear terms the broader scope, the connecting links and contingencies of reinforcement that contributes to the perpetuation of the deprivation of Blacks in America.

Could it be that the roots of the problem have always been economic greed rather than racial preference? Could it be the perception of Blacks as children on the part of Whites that contributes to the view of inferiority and unequal status? The relation to acquiring resources, managing and conserving resources is ultimately related to power.

Could it be that the root of the problem is the separation of power potential and power sources from the Black community?

This book skillfully describes the underpinnings of the system that creates "trap doors" to insure that Blacks hang themselves with their own weight.

The relation Blacks in America had with Africa and the relation Blacks have with Africa today are analyzed in a concise way as to draw clear conclusions. The untapped resources in Africa causes one to wonder whether the underdeveloped countries in Africa are considered to be "never to be developed" countries in Africa. Should the resources be used to make a few more people wealthier or should they be used to benefit the masses? Should the top be lowered or should the bottom be raised?

The book ponders the most important issues of our time and explores the theoretical as well as the practical application of concepts that can indeed lead to a matter of Black transformation - concrete solutions to the problems of Black America which will indeed impact global solutions, for our destinies are tied together.

"Injustice anywhere is a threat to justice everywhere," Martin Luther King Jr.

Dr. Bernard LaFayette

Director, Center for Peace Studies, University of Rhode Island

Preface

Since Alex Haley's magnum opus *Roots* established the historical truism in 1977 that Black people in America are Africans, and showed the bitter and painful journey in the slave ships across the Atlantic to the new world, the connection between motherland Africa and Black America has mainly been reflected in abundant talk and little action.

That is why I wrote this book and called it *A Matter of Black Transformation* because, looking at the conditions of Blacks in America and their counterparts in Black Africa and other parts of the world, it is evident that Black people around the world are caught up in the same socio-economic struggle. To our collective chagrin, we are facing the same fight for real meaningful education, economic empowerment through enterprise development and true political liberation that gives people effective political participation. We have omitted from the table of hope the challenge of addressing problems of less-empowering education, economic subjugation, social alienation and grand political deception.

Today in the Black world, we have institutions that boast of longevity but little relevance when viewed against the backdrop of improving the dignity and living conditions of our people. They regrettably exist to boost the ego, public visibility and financial strength of their top brasses, and not for the greater good of the greater number of our people.

An assessment of the level of poverty in Black communities around the world, and how we are missing the economic empowerment train in an era of globalization triggered in me a great desire to write this my second book to stir us up to truly redemptive actions. As a Black journalist, I have seen first hand the suffering of the masses of our people. I saw it in Black Africa and have been seeing it in Black America. I have seen the devastating result of a lack of meaningful education that leaves Black children with no real sense of identity and eventually makes them vulnerable and endangered. I have smelled the stench of racism masquerading in our public sphere as formalistic impersonality. I have also witnessed the blatant hypocrisy of some who claim to be the social liberators of the Black world.

Sometimes, I am tempted to think the so-called struggle for Black liberation is nothing but a game. And this game, it appears, is played by members of the elite class at the expense of the masses of Blacks. But then, I quickly realize that no matter how Black liberation is misused for self-aggrandizement, we should not allow ourselves to be distracted from the real struggle to transform our situations, to set ourselves free from our present socio-economic and political menace and to take charge of our own destiny.

Transformation can only come about through practical, result-oriented ideas and actions. Enough of the long talk! Let's do something visibly redemptive.

Throughout this book, you will find practical things people in Black America and Black Africa can do to bring about the much-needed transformation. For instance, I recommended that Black American businesses should be involved in the economic and political life of Black Africa. There are many White transnational corporations operating in Black Africa while Black American businesses are sleeping. I believe it is not enough to connect with Black Africa at the cultural level and ignore the economic level. Black Africa and Black America must connect at all levels for the betterment of the Black world.

I talked about the need for an African-centered educational approach that helps Black children gain a genuine sustained sense of cultural identity. I am aware that many people before me have raised similar issues in books, but I hope this work will be an important and stimulating addition to the pool of knowledge on the subject; thus increasing our awareness and consciousness to enable us leap forward into the future.

However, I did not just sit down and begin to pen this book. Apart from reading lots of books, I had conversions with a number of people to get their viewpoints on the issue and many were concerned about the state of the Black world, which is why you will find appendixes towards the end of the book. These are interviews with relevant people on issues like Black education and Black enterprise development, which I consider very important to you, my reader.

I also had exchange of ideas with wide-ranging Blacks, some of whom I had met at seminars and conferences, and others in the course of my work as a Black journalist and public lecturer.

In all of my opinion surveys, I found college students to be the most concerned about the future of the Black race in an era of globalization. At Wayne State University, I faced questions during many guest lectures I gave at the Department of Africana Studies on why the relationship between Black America and Black Africa is not as strong as it should be.

Why is the Black world in such turmoil?

In this book, you will realize that lack of proper Black education is to blame for most of our woes. We do not teach our children (and even adults) who they are, where they are coming from, where they are, where they should be, and what they can do to get there.

I identified the reasons for some of the problems we face in Black America, showing, for instance, the consequences of our lack of understanding of

Black Africa and how it has contributed to our "non-existent knowledge" of the continent. The "Africa is garbage" image festered in the mainstream media has seriously distracted the attention of many Black Americans and thus affected the relationship between Black Africa and Black America.

Such image, for instance, is largely responsible for the inaction of Black America in the face of political problems like the Darfur genocide in Sudan and child exploitation in Liberia, which I examined in this book.

Black America has an annual purchasing power of about $700 billion which should be translated to wealth for Black America and by extension Black Africa through vigorous enterprise development and consequential gainful employment creation and improvement of the living conditions of our people. Thus, there is no reason for poverty to ravage our communities. There is no need for the Sword of Damocles to continue to hang over us.

Things will change for the better when we all – big or small, in Black America or Black Africa - play our parts consciously and with resilient commitment.

Bankole Thompson
bankole@bankolethompson.com

Chapter One
Pushing frontiers of Black education

In an era marked by a plethora of buzz words, one may be tempted to see Black transformation as yet another literati buzz word. Thankfully it is not. Black transformation is about changing our conditions, our lives. It is about our self-knowledge, self-direction, self-motivation, self-correction and self-restraint. It is about breaking the barriers of complacency or resignation and marching conscientiously towards better life and better community with the power of relevant, contextual education, among others.

I often talk about Black education as the engine room of the new Black society we desire but need to exercise our hearts, minds and hands diligently and steadily to accomplish.

But Black education is not acquiring college certificates as some think it is. Rather, it is about knowing who we are, where we are, where we are coming from, where we should be and how to get there. It is about developing the resilient capacity to be in control of our challenges and situations in the Black world.

And perhaps, nothing opened our eyes to who we are and where we are coming from more than the movie and book *Roots* written by Alex Haley. An instant classic in the Black experience, *Roots* provided Black people around the world a veritable platform and avenue to view their authentic history and culture.

Remarkably documenting the history of the Black American slavery experience, Haley forced a new expository and educational discussion on an issue that is not often given the prime time opportunity on American national television and TV screens across the world.

Roots became a wakeup call to Black Americans to see slavery as not only a historical genocide, but also a source of strength from which to draw courage and determination in charting a new redemptive future.

The vivid portrayal of Black Africans revolting on slave ships against their slave masters projected a historical truism that Black Africans had indeed strongly resisted dehumanization and exploitation on the continent. *Roots* made it clear that our Black ancestors had launched a strong resistance movement. They had fought the inhuman treatment of the enslavers and physically demonstrated their disapproval of the gross human rights violations the so-called civilized world had unleashed on Africans. Even when a number of them were thrown into the sea as the painful voyage to the new world continued, our ancestors did not relent or relinquish their struggle.

So *Roots* laid bare for us the significance of our origin and survival thrust, and that strong sense of origin and survival is greatly needed today as Blacks in different parts of the world continue to deal with bigger issues surrounding globalization.

But *Roots* had also revealed a deeper understanding of the African society that is hardly brought to the fore. The movie featuring popular Black actors like John Amos who has a deep understanding of Africa, showed that the Black experience did not begin with transatlantic slavery.

Haley, who traced his roots to the tiny West African country, The Gambia, brought to the fore the fact that Africans had vibrant, well-organized societies before transatlantic slavery.

Creating and preserving our institutions

I have visited the Holocaust Memorial Center in Farmington Hills, Michigan described as America's "first freestanding museum, dedicated to the memory of the Holocaust." Walking in the museum, the images from the Holocaust provides any visitor with a vivid reminder of what took place in Adolph Hitler's Germany decades ago.

"The tale of the Holocaust is one of epic proportions. It is a history of evil, although the evil is undeniable, but also of great courage, strength and righteousness," HMC said on its website.

Looking at such a magnificent educational and history-preserving institution created and maintained by the American Jewish community, I wondered why the American Black community whose experience goes back as far as 400 years have not yet established a similar institution that would help our children understand the Black holocaust in its essentials.

Detroit is home to the largest Black museum in America. Called the Charles Wright Museum of African American History, the museum has not been utilized enough as an effective educational tool for the Black experience. It has caught a lot of political flak with arguments on whether it is self-sustaining; questions about its fundraising ability and the competence of the staff with regard to increasing membership.

Such questions came to the fore in 2004 when the museum announced that it would be closing shop unless it received $2 million to keep lights on and finance day-to-day operations. It had a cash flow issue and could not meet payroll. News of the museum being in dire straits surprised many in the Detroit community. How could an educational institution of such magnitude operate without an endowment fund? Where were the marketing strategists to come up with creative ideas to draw people to the museum? Does the museum have an effective public relations arm? Questions upon questions, one may say, but they certainly needed to be asked.

But for the never-say-fail spirit of US Sixth Circuit Court of Appeals Judge, Damon Keith, a Black American, the museum, perhaps, would still have been in financial limbo. Or worse still, it could easily have closed down to our collective shame. After receiving the bad news about cash flow problems at the museum, the Black jurist summoned museum board members as well as business and political leaders to his chamber in downtown Detroit and challenged them to demonstrate support for the largest Black museum in America.

A one-of-a-kind museum telling the story of the Black experience through exhibits, audiovisual presentations and events, the Charles Wright Museum of African American History provides a great reservoir of knowledge for Blacks. Those who were present in Keith's chamber said the respected jurist handed a charge to his guests and questioned the wisdom of sitting by and letting the museum go down the drain.

However, the problem with the Detroit museum proved a larger issue in Black America: Our shaky commitment to the educational and cultural institutions that speak to our identity.

Sure enough, the Detroit museum was not the first to be in such dilemma. In fact, some Black colleges are facing similar financial hardship. Some have been closed down because they defaulted in their federal loan programs. Defaulting in federal loan programs for three consecutive years could cause an institution to lose its federal aid status and losing its federal aid status could have other serious implications for a Black college even though a 1998 legislation allowed Black colleges who are in default to continue in the federal loan program by filing a petition with the US Department of Education.

I believe the fact that Black schools default on loans in the first place speaks to the wider issue of poverty in the Black community. A survey of 14 Black colleges across the country showed 52 percent of the students came from families earning less than $20,000 per year. And more than 50 percent of those who defaulted on loans came from families earning less than $10,000 per year.

The colleges that were surveyed were Allen University (South Carolina), Arkansas Baptist College (Arkansas), Barber-Scotia College (North Carolina), Central State University (Ohio), Houston-Tillotson College (Texas), Jarvis Christian College (Texas), Lane College (Tennessee), Mary Holmes College (Mississippi), Miles College (Alabama), Paul Quinn College (Texas),

Southwestern Christian College (Texas), Texas College (Texas), Texas Southern University (Texas) and Wiley College (Texas).

Black Excel, an online educational site, named as a major college help resource by Ebony Magazine and Black Enterprise in 2006, released a statement on its site when rumors had it that the 14 Black colleges listed were about to close.

"As has always been the case, our colleges are often struggling on tiny endowments and sometimes are in a state of financial stress. For this reason, we at Black Excel strongly believe that we should be putting more money into the support of our schools. We must take care of our own because we need every advantage we can muster," Excel stated on its site.

And that's absolutely true!

Teaching Black children entrepreneurship

Our educational system in Black America, an extension of the one in mainstream America, basically teaches Black children to acquire college certificates, get weekly or monthly low paying jobs, and start depending on loans for needs like housing, house furnishing, car, health care and personal or family vacation.

Clearly, the notion of going to school to earn a degree that can get one a job is not working especially when outsourcing is displacing many Blacks in the marketplace. When you dial a Detroit Michigan phone number and someone answers you in India, you don't need a globalization expert to tell you some jobs have been shipped out from Detroit to India.

If very many well paying jobs are leaving America, what should Blacks be doing? Lamenting and inviting pity or actively creating our own well paying jobs? The latter, of course! Therefore, our educational institutions must begin to teach Black people the good old friend called entrepreneurship to enable them create their own jobs – like web 2.0 businesses - even while in college. We can't continue to depend on an educational system that is geared towards certificates and paid employment, when we can have one that offers concrete opportunities for decent careers and life-enriching self-employment.

In other words, enterprise development should be a major focus of redefined education for the

Black child in America. It is past time for Blacks to simply show a laundry list of employment history but a no-show in enterprise development and property ownership.

In the American public school system, there is unequal funding of public schools in Black cities and their counterparts in rich suburban White neighborhoods. That, in itself, is a challenge for Black institutions with the resources to start concrete educational programs that do not only address the issue of growing unemployment in Black America, but also change the underlying mindset of going to college just for the purpose of getting a job.

In order for such an educational system to succeed, Blacks need to take ownership of these institutions in their communities, and there should be clear vision and willingness on the part of our educators to see this through.

Let's have African-centered education

Some schools in Black America are already implementing African-centered education in their curriculum, but how I wish it is all of them. African-centered education, which embodies a study of the culture of African people, is needed for Black children in America to know their past, appreciate their present and shape their future.

If Black America is to experience a leap towards a meaningful education, providing opportunities for Blacks to learn about Africa and its socio-political and economic life is key. There is no way you can expect a Black child to grow to accept Africa, with all the stigmatization of the continent, without an educational thrust that reveals Africa in all its essentials.

Africa is not a country as some people think. It is a continent of about 11,668,545 square miles and 54 countries like Nigeria, Ghana, Ethiopia, Kenya, Sudan, Gambia and South Africa. How many Black American children know that basic fact? Few, I believe. But how would they when many Black American adults don't know much about Black Africa other than as the origin of their ancestors or as a place filled with HIV/AIDS, malaria, civil wars, child soldiers and poverty – just as the mainstream media would like them to see it?

Instructively, the University of Iowa in the largely White state of Iowa runs a well attended summer camp called *Africa is not a country* for school children in that state. Are universities in largely Black cities like Detroit and Atlanta creating such awareness? No!

Africa is a natural and human resource-rich continent of about 740 million mostly Black people, with some Whites in countries like South Africa and Arabs in countries like Algeria, Libya, Egypt, Morocco, Tunisia and Mauritania. That should be taught to Black American children from as early as age 6 – at home and at school.

Black educational institutions should be providing the avenue for their students to frequent Black Africa (comprising 44 countries in Sub-Saharan Africa) and relate better with its peoples. In other words, there needs to be more Black American students going to do internships and fellowships in Africa. After all, a good number of White students go to the continent for fellowships and internships. Besides, Washington DC has Africa advocacy groups with White student interns, and this, in a way, challenges Black students.

In June of 2003, I left Detroit for Washington DC on a foreign policy fellowship to work for the Africa Faith and Justice Network (AFJN) as a visiting journalist. AFJN is one of the leading advocacy groups in Washington that promote a better US policy toward Africa.

In the early 1990s, AFJN was the organization that first brought the Liberian crisis to the attention of Washington DC. It led the discussion that provided a deeper understanding of the crisis in Liberia and helped to generate a response that benefited the people of Liberia.

I had visited Liberia in 1999 to attend a communication conference of media practitioners and international policy makers. Throughout my one week stay in the West African nation, I saw partially destroyed buildings with bullet marks on them. I saw Liberian children that had been indoctrinated or rather coerced into child soldiering during the civil war.

On every check point, child soldiers were the ones giving directions to cars entering in from Robert's Field, the Liberian airport. In broad day light as I drove around to see the devastation of the war, I saw young soldiers scrambling for food rations distributed to them by the government of brutal dictator Charles Taylor who was governing the country. The taxi driver let me know it was a politically volatile situation as the young soldiers, after fighting the senseless war, had little or nothing to eat. Failing to provide food for them could mean risky political climate Taylor was not willing to tolerate under a regime where people lived in fear and democracy and the rule of law existed only in the realm of imagination.

So, when I got to Washington DC as a visiting journalist for the AFJN, I appreciated the work of this organization in helping the people of Liberia. The day I reported to the offices of AFJN to commence my fellowship, I was introduced to the staff. And before long, I found out that the West Africa desk on conflict resolution which I was going to head throughout the length of my fellowship was previously run by a White female student at John Hopkins University. The student, according to Dr. Marcel Kittisou, AFJN director at the time, had gone to Guinea Conakry in West Africa where she was doing her Master's dissertation. The information did not sink in until I discussed with a couple of friends about how Whites have taken over the Black Africa fight in Washington DC.

Most of the institutions inside the Beltway advocating for just and humane policies toward Black Africa are run exclusively by Whites who frequent Black Africa. That is why you can tune in to CNN and all you see are White analysts explaining to Black America what is happening on the African continent. That however does not undercut the role of some Black Americans who are deeply engaged in some of the issues that are important to Black Africa. Organizations like Africare aptly described by venerable international statesman Nelson Mandela as "America's gift to Africa," are working tirelessly to help the African continent.

Founded by former Africa director of US Peace Corps, C. Payne Lucas, Africare today stands as one of the foremost Black American organizations involved in the life of Africa. But Africare should not be the only organization that Blacks in America can dangle to show their commitment to Black Africa. It should not be the only institution that should provide education about the continent from a more accurate, well informed and reliable perspective.

In April of 2006, the University of Iowa brought me to campus as keynote speaker for the university's Africa Week celebration. The university, through its African Students Association, requested that I speak on the theme "The Media and the African poverty crisis." Everyone I spoke with in Detroit joked about my visit to Iowa because it is a conservative state.

"What are you going to do in Iowa?" was one of the questions a friend of mine had asked. I retorted that the university had invited me to speak during their Africa Week celebration. It was unbelievable to my friend that a White university like that would devote a whole week to Africa and highlight the continent, dealing with the contemporary issues it is facing. But the bottom line for the university was that it was an educational venture. The university wanted to instill in its students a global education, which meant studying the continent of Africa.

After my well received Friday lecture came next day's dinner and awards night at which the African Students Association honored faculty, staff members and students who had shown strong commitment to Africa. The night was an enjoyable one for me. Not only did I take a break from my busy journalistic assignments in Detroit, but I was able to visit the land of the corn field and see a different socio-economic climate.

My colleague who lives in Iowa with his wife jokingly told me when I landed not to expect a civil rights conference given the state's pro-business thrust. But while civil rights issues were not the primary reasons of my visit, Africa had to take the center stage. On the evening of the awards, White faculty members spoke about their sacrificial educational and humanitarian services to Africa over the decades. Some had lived and taught in Nigeria and others had worked in other African nations.

Beautifully clad in their African costumes and offering some kind words in African languages, they spoke passionately about the continent of Africa and how they have practically identified with the physical needs of indigent families and communities on the continent. I sat down soaking it all in and thinking rather understandably about Detroit and what many Black colleges and Black cities are not doing. When I got back to Detroit, I shared my experience with an Africana studies class at Wayne State University and even suggested to them to celebrate Africa Day every year on campus, through their student government council.

No doubt, a majority Black city like Detroit should not hesitate to have such a cultural cum educational fanfare that could help inspire us on the path of Black transformation.

I believe there is no excuse for a Black child in America in this 21st century not to be familiar with at least leading African literature from notables like Chinua Achebe, Wole Soyinka and Ngugi wa Thiong. The same is also true of African students on the continent who should be familiar with the works of Maya Angelou, Toni Morrison and Walter Mosley.

Promoting education between Black America and Black Africa should be a two way street. Young African students should be conversant with Black American literature because through such a medium, they are able to understand the culture and challenges of Black America, and perhaps put themselves in the shoes of their Black American brothers and sisters since the first slave ship landed on Annapolis.

One other important thing is for Black America to start recognizing major African holidays like June 16, which is commemorated every year by the United Nations and the African Union to remember the 1976 massacre of 700 Black South African students.

Wondering why the massacre? The young students were demonstrating against the very repressive educational conditions imposed on them by the White racist apartheid regime of South Africa. They wanted good education, one of the best gifts every child greatly deserves.

The Day of the African Child, as it is called, puts the courage and sacrifice of those children in meaningful perspective while highlighting the plight of today's African children in war-ravaged places like Darfur Sudan and the mineral resources-rich Democratic Republic of Congo.

There is no reason for Black America not to commemorate this holiday and use the Soweto massacre as a backdrop for looking deeply at the condition of Black children in America today with a view to responding adequately to their needs and preparing them for the challenges of the present and the future.

Black education in the Internet age

The Internet is one of the most effective and pragmatic tools in fostering educational exchanges and cooperation between Black students in America and their African counterparts.

Sitting in a class at Wayne State University, a student can read what another student in The Gambia, Kenya or Sierra Leone emailed or posted on a college website or have free live exchanges of ideas with Black African students by video chat or with Skype.

Prior to the Internet, we wrote letters, which took days before reaching their destinations. But with the Internet, Black students can easily exchange useful information on their research projects, research results, and issues and events happening around them. Therefore, Black teachers and educationalists should encourage Black students to establish and maintain online communication with fellow students in Black Africa, the same way White students communicate with fellow students in Europe and Asia. Black students don't have to wait on NBC or any of the national networks to report on a story to satisfy their curiosity about Black Africa.

Through the exchange of emails and daily or weekly Internet contacts through discussion boards, social networking sites, text messaging, video chats and others, they can reach the primary sources of the news they see on television. Is the Internet available in Africa? You asked. Yes, of course! Countries like South Africa, Nigeria and The

Gambia have a good Internet penetration. Are there students having great time in Africa, enjoying comfort and pleasure? Certainly! Is everyone in Africa poor? No! Does Africa have modern cities? Visit Abuja and Abidjan and you will be surprised. Does Africa have celebrated great achievers? Yes, in sports, cinema, literature, community leadership, academia, research, medicine, diplomacy, inventions and so on. Africa has its own fair share of achievers many people don't hear about in the Northern Hemisphere.

Yet, many Black students or Black parents in America shy away from visiting Africa just because the news media regularly paints nasty pictures of Africa. I believe very practicable, regular Internet contacts with Black African students in Africa can help counter the negative-only news that emanates from Africa. This is especially because education should confer on one the ability to make a critical judgment of issues, to arrive at conclusions based on deep-seated review of issues or situations, conclusions based on proper, adequate information that nourishes the curious mind.

The other point that needs to be made is the increasing number of Asian students who have dominated the technological training marketplace. In almost every tertiary institution in America, Indian, Chinese, Japanese and Korean students have taken over the computer science and engineering disciplines which prepare them for high paying jobs or profitable self-employment in any part of the globe.

Black parents, policy makers and educators must therefore direct their focus to sciences and technical education to enable Black youth compete favorably in the global marketplace.

They cannot just fold their arms and watch as the Asian technological giant, India, has become the largest source of foreign students in American universities. A report in 2002 released by the Institute of International Education (IIE) revealed that 67,000 Indians enrolled in US institutions, surpassing China, which dominated the field for much part of the 1990s.

Chapter Two
Help! Black economic power in chains

The widening gulf of poverty and social alienation we see around us today point to the blazing challenge of Black people around the world to think and act in pursuance of meaningful transformation and cease relegating our energies to the deepening valleys of lamentations and excuses. We need to start climbing the tough but scalable mountain of economic empowerment by doing practical, non-exclusive, result-yielding and situation-changing things that capture the realities, dynamics and intricacies of our local, national and global economic landscapes.

No day passes without us hearing legitimate complaints about the bad things others are doing to us – the stench of racism, the heartless acts of dehumanization, the bouts of exploitation, the sting of deprivation and the menace of extortion, *inter alia*. Despicable as such bad things are, the point must be made that there are also unacceptable things we are doing to ourselves in the Black community - the self-inflicted wounds of family decimation, the mind-boggling passivity before yawning nets of opportunities, the ubiquitous and misery celebrating pull him or pull her down syndrome, the stay-on-top-and-keep-others-under thrust, and the wrong-premised, disjointed and dehydrated deeds of negligible results that tend to stagnate and frustrate us.

Put differently, it is not enough to lament, proffer excuses or apportion blames. We have been inundated by all that. The time now is for positive result-driven actions.

If we in the Black community have about $700 billion annual purchasing power as some statistics say, one wonders why we have not been able to turn such huge purchasing power into real economic power by creating much-needed meaningful jobs in the retail, manufacturing and service sectors and taking millions of Blacks from the morass of under-employment to the savor of sustainable gainful employment.

Why do we sit down and watch entrepreneurs from other ethnic communities - like Asian businesses in America that dominate the multi-million dollar Black hair product market - take full control of various aspects of our economic lives? And how do some of our leaders react to it? They blame the non-Black entrepreneurs for being bold and perceptive enough to come to our backyards, identify our very obvious needs which we are not responding practically to, and consequently make wealth providing us with such needs. Yes, they blame the opportunistic, barrier-defying, globalization-minded entrepreneurs, instead of our tendency to rigidly tie our well being and future to the vicissitudes of partisan politics while paying lip service to the practical challenges of economic power.

Such blame game crumbles in the face of invaluable critical and strategic thinking, which seems to be in short supply among us or is, simply put, treated with levity today as we grapple with the numerous issues and situations confronting us.

It bothers me that Black America and Black Africa have high-powered organizations with the socio-political and economic clout to make wide-reaching positive things happen in our communities but are not making a measurable difference, or are they?

Only very few are rising to the occasion, I must state. The others unfortunately exist only in name and signpost, and confine their significance to elaborate black-tie, long speech, and designation and connection parading events primarily aimed at spotlighting the leaders and their privileged families and circles of friends. Though such events may pretend to be concerned about the living conditions of the generality of our people by talking about fundraisers, they usually serve the power play needs of some members of the elite class or some wannabes. That's why I have said at many speaking engagements that if there is a membership organization out there that has shown it has outlived its usefulness by not responding to the pressing needs of Black people, it should helpfully be sentenced to the dustbin of history where it belongs.

There are many membership organizations that host black-tie events but cannot show concrete ways they have positively changed the lives of the masses of our people and they should no longer be seen as representing or speaking for Black people. They are representing and speaking for the fat-pursed ones in the driving seats, not the suffering masses of our people many of whom are unemployed, underemployed, malnourished, unable to pay for college, facing foreclosure or sick without medical coverage.

If these organizations are set up to liberate the Black community from political and especially socio-economic subjugation, they would be led by sincere men and women who understand the dynamics and challenges of contemporary global economy and the virtue of leadership and personal sacrifice.

Organizations who claim they are working for Black people must be led by individuals who are not as much committed to accumulating wealth and political heritage for their family members as they should be to stimulating collective well being and progress and speaking for the downtrodden and defending the defenseless.

Black America longs for leaders with the appropriate recipe for transforming our $700 billion annual purchasing power into meaningful wealth for all segments of our community. After all, we live in a world society where an individual's social relevance and influence are largely defined by how gainful their relationship is to the means of production.

America is a capitalist society and therefore the economy is run by captains of industries and the entrepreneurial class, not the underclass. So, we need to greatly increase the number of Black-owned companies, Black captains of industry and Black middle and small scale entrepreneurs, if we are going to be serious players in the new borderless capitalist marketplace.

Wondering why many politicians don't buy into our Black agenda despite our publicly demonstrated legitimate discontent? It's because he who pays the piper, as the cliché goes, dictates the tune.

Black and White politicians alike are typically bankrolled by captains of industry and the entrepreneurial class, so they obey them and disobey us, even though they are morally and constitutionally obliged to obey us since we voted them into office.

It therefore serves us no purpose to continue to extol big name organizations and leaders that are bankrupt of ideas when the need is truly moving our community forward. It is dishonest or at best naive to lead people under the sole refrain of racism, while putting economic empowerment in limbo.

Leadership in Black America today can be best described in what Bruce Dixon, managing editor of the Black Agenda Report, wrote about Atlanta's poverty:

"The contrast between the prophetic leadership of the civil rights movement era and the profit-oriented leadership of the Black business class that came in its wake could not be clearer. Old Testament prophets, like civil and human rights advocates, were not millionaires or kings," Dixon wrote.

The Black American challenge

There is no gainsaying the fact that Africa is the richest continent in the world, endowed with many of the globe's finest natural resources. However, Africa has received so much bashing in the national and international mainstream media, which deliberately presents the continent as a place unworthy of any sort of foreign investment. That is what I call the "Africa conspiracy," driven by the mainstream media, which continues to perpetuate the image that Africa is a dark continent.

It is very evident that the mainstream media is playing to the gallery of the hegemony- driven Western governments and transnational corporations. While the mainstream media machine of Western nations like Britain and America continue to beat the drum that Africa is a scary place to be, multinational corporations are investing heavily in the continent's oil and mining sectors, among others. If Africa is so bad, why are companies such as Leon Tempelsman & Son, a company active in mining, investment, business development and minerals, operating in Africa.

But they are not the only businesses investing in Africa's vast natural resources. There are the Chevrons and the BPs, for instance. Multinationals abide wherever natural resources abound in Africa. They exist and operate even in times of bloody wars and Darfur-like genocide. But where are African-American businesses that could be more humane and truly development-driven? Where are African-American entrepreneurs with cognate ties to Africa?

In Washington today is situated the Corporate Council on Africa run by a prominent Africa consultant inside the Beltway, Stephen Hayes.

Founded in 1993, one of CCA's many programs is The Meetings Program, which "regularly holds seminars with visiting African ministers, members of the African private sector, US Administration officials, members of Congress, US and African Ambassadors, and representatives of various multilateral agencies. CCA coordinates private meetings for member corporations with senior African and US government leaders."

According to CCA, in 2006 alone, the group arranged meetings between 34 African heads of state and its membership. Thus, while mostly White corporations were busy connecting and building partnerships in Africa, African-American businesses were apparently detached.

Where is Black America's equivalent of CCA? Certainly, we do have focus groups that help several causes on the continent, but there isn't any strong chamber in Black America that is seriously connected with Africa and doing what the CCA is doing with the continent. And that is largely due to a lack of understanding of contemporary Africa. Most Black Americans still see Africa in the image that was created by institutions and individuals desperately seeking to distant Black America from Black Africa.

The ugly and baseless notion of Africa as empty and bestial still exists. However, with the growing number of well educated African professionals visiting or immigrating to the US, and African studies becoming a curriculum in community colleges and universities across the country, many African-Americans are seeing and feeling the need to engage more with Africa.

But the idea and practice of businesses and leaders from the continent hosting or featuring in major events that illuminate the continent's numerous potentials is still in its infancy. Ivy League universities may sometimes invite leaders from various sectors of the African economic and political life to give lectures, but such lectures are usually confined to only those students who could afford to attend Ivy League colleges.

The challenge therefore should be for colleges and universities in urban centers like Detroit, Chicago and Atlanta with high concentrations of Blacks to organize regular public seminars that will spotlight Africa in its totality as part of their African studies programs.

That would mean bringing in resources from the continent to do so. Why not? If Yale and Harvard and other universities find such revelatory and intellectually stimulating engagements essential to their research and learning, colleges and universities in Black cities should treat them as such. If the leadership of these institutions is willing to expand the frontiers of their educational pursuits by re-examining Africa and its relationship with Black America, they could herald the desired Black community of greater and wider-reaching opportunities and capacities that liberate socio-political and economic dignity from the exclusive preserve of the elite class.

The local Black chambers of commerce cannot be divorced from this reality either. In fact, the chambers are often more positioned to facilitate such partnerships in an effective way because their memberships comprise established Black American businesses fairing well in the American marketplace. But there has to be willingness on the part of these local business leaders to reach out to Black Africa.

It is not enough to talk much about being connected to Black Africa by ancestry only to sit passively and watch White corporations mine the continent for fine gold as if it is a Whites-only affair.

Chapter Three
Some Lessons from China

At the global level, China is courting the continent of Africa and its leaders into doing extensive business and creating strong wealth-creating partnerships. Yet, Black America mainly thinks of Africa in terms of cultural gratification instead of economic possibilities. Black America has to wake up. Whether we like it or not, China is in line to become the next undeniable super power on the world stage. Chinese businesses are infiltrating every corner of the world – even in the Northern Hemisphere.

Chinese restaurants, for instance, are in every major city you can think of on the face of the globe. Most people around the world today wear clothes and shoes made in China just as most children use toys made in China.

China scored a major influence with Africa when it organized a meeting in Beijing in 2006 with African leaders to discuss trade and other bilateral issues. That should not be shocking to anyone because China sees Africa as a great business partner and respects that business linkage.

According to the influential American think-tank, the Council on Foreign Relations (CFR), China's trade with Africa in 2006 reached $50billion. This was seen as a welcome development by most African nations even though China has come under fire for refusing to pressure

Sudan on the Darfur genocide because of its oil interest which will be dealt with in a subsequent chapter.

Some who are observing China's relationship with Africa are left to wonder how this began. Deborah Brautigam of American University is the author of an insightful book, *Chinese Aid and African Development: Exporting Green Revolution.* Brautigam contended that China's involvement with Africa is not a new venture. It is not something that started overnight as some are made to believe. Instead, it started in 1979.

"During these years China kept up an active menu of aid projects in more than forty-five African countries. Their annual aid commitments in Africa sometimes surpassed those of Japan, Norway, Sweden, and even Britain. The flurry of activity we see today has deep roots," Brautigam said during a February 2007 online discussion on the CFR webpage. Brautigam gave many examples of China's investments in Africa, showing the extent of the world's next super power's commitment to doing trade with Africa.

"In Angola, China's recent $2 billion and $2.4 billion Eximbank credit lines were tied to infrastructure investments. Teams of Chinese are already in the country building roads, rehabilitating railways, and building schools and a huge neighborhood of low-cost housing. Thirty percent of the contracts under the loan are targeted to Angolan firms. Angola pays for this infrastructure with oil.

"Compare this with the completely non-transparent $2.35 billion loan extended to Angola by Britain's Standard Chartered Bank, Barclays, and the Royal Bank of Scotland. China has also been proactive on Africa's debt burden. They regularly cancel the loans of African countries, loans that were usually granted at zero interest. They do this without the long dance of negotiations and questionable conditions required by the World Bank and the International Monetary Fund."

So the evidence that China is strongly investing in Africa is overwhelming.

Louis James, president of JASCO International, a Detroit-based African-American owned global automotive supply chain management and logistics company, believes not much is being done on the part of Black America to trade with Black Africa.

Louis was a member of Wayne County (Detroit-Michigan) trade mission delegation to China in 2006 led by county executive Robert Ficano.

"Our goal in China was to develop a relationship with the Chinese manufacturers, encouraging them to relocate to Wayne County-Detroit. The relationship building shall be with the Big Three (General Motors, Ford Motor and Daimler Chrysler) and the manufacturers that we meet in China. As a result of the trade mission, JASCO International was able to meet with over 200 companies and to gain a competitive edge with worldwide alliances.

"The efforts of Wayne County and Mr. Ficano during this trade mission are commendable and highly regarded by JASCO International and the team of local business people who attended. In this competitive automotive industry, it is very important that local companies seek to do business oversees and the Wayne County trade mission was a venue that gave us access to numerous companies to pursue those efforts," Louis said in a release about the ten day visit.

Interestingly, it was during a 2007 Chinese business conference at the Ritz Carlton Hotel in Dearborn Michigan that I knew about Louis' visit to China. We got into discussion about China's strategic moves in the global marketplace and how this soon-to-be super power is displacing some of its rivals in the market by reaching out to nations across the globe. Soon, our discussion changed to Africa, especially because I knew Louis and his brother, John, have shown strong commitment to Africa by sponsoring various projects in some countries.

One of such countries is The Gambia where John spent thousands of dollars in the West African nation helping erect irrigation projects and other essentials in villages. So, during our conversation at the hotel prior to the start of the conference, I told Louis I was working on a book on Black transformation and that several chapters in the book make the case for exploring economic opportunities between Black Africa and Black America.

If China and Chinese businesses are in Black Africa, why not those who are connected to that part of the world by blood and ancestry? Louis not only agreed with the premise of my book, but also expressed willingness to share his experience in

China and how Africa was given the utmost respect it deserved as a key player in the international political and economic scene. He invited me to his office to talk about Africa and China and the challenge for Black America, especially against the backdrop that Africa was given the respect it deserved in Beijing in 2006 when the continent's leaders visited at the invitation of China.

"I watched the news while we were in China, and saw Black limousines taking African presidents around. The whole of Beijing was literary shut down. The Africans were treated with respect. I'm sitting in my (hotel) room watching how the Chinese were working this - treating Africa as a respected business partner. It was an eye-opener for me," Louis told me during an interview at his Fort Street office in Detroit. "China treated their African delegation the way heads of state are supposed to be treated."

Talking to Louis, I could see a passion and willingness in him to reach out to Black Africa. He said the reception Beijing gave to the African delegation was something unheard of in America because of racism and prejudice. For a minute, I thought the savvy businessman I visited was turning to a Black activist. I thought I was in a class on Africana studies or international relations.

"Because of racism and prejudice in America, when African leaders come here, there is no fanfare," Louis said. "And the people who are aligned with them during their visit have no clue about the need to connect with Africa."

In Louis James' view, there is a reason for this. "There is a purpose for this - for Black people not to create a kingship with Africa. We have to become part of the African fiber. We probably have more appreciation for others than for Africa. This is the richest continent in the world."

Thus to Louis, the China-Africa experience was a strong wake up call for Black American businesses who must be driven by a consciousness that Black Africa must take its rightful place in the African-American experience. American Blacks, he further pointed out, must see their development as tied to the growth of Black Africa. And both must come together in forming a strong linkage that goes beyond mere rhetoric and occasional tourist visits.

If education, as Louis put it, must be the first place to start, Black American businesses must first express a willingness to be truly educated about Black Africa instead of relying on books that misrepresent the motherland and push one of the world's greatest lies that Black Africa is dark, empty and worthless.

Black American businesses must see themselves as an extraction of Black Africa. With that mindset, it becomes easy to find the way to honest and serious dialogue and socio-economic cooperation based on mutual respect.

Such a paradigm of doing real business with Black Africa, according to Louis, should be promoted by the Congressional Black Caucus (CBC).

Besides, I think Black America needs to have its own version of the CCA that would serve as an effective economic engine in fostering dialogue – in the form of business conferences and workshops – business matchmaking and collaboration with the Black African business sector.

Chapter Four
Whither African Growth and Opportunity Act (AGOA)?

African Growth and Opportunity Act or AGOA, as this act is commonly called, was introduced on May 18 2000 during the administration of President Bill Clinton to enhance trade between the US and African nations. After a five-year battle in Congress, Clinton finally gave his assent to the legislation that would make it easy for Sub-Saharan African goods like fish, textiles and cut flowers to be imported and sold in the United States which has a market of over $11 trillion.

Many lawmakers at the time hailed AGOA as a legislation that "advances US economic and security interests by strengthening the relationship with regions of the world that are making significant strides in terms of economic development and political reform."

It has since then had three amendments in 2002, 2004 and 2006 under President George W. Bush, aimed, as it were, at making it work better.

Thus by 2007, the United States Trade Representative, Ambassador Susan C. Schwab, was reporting to Congress that the US "devoted $394 million to trade capacity building activities in Sub-Saharan Africa," in financial year 2006 which is up 95 percent from financial year 2005.

The website, AGOA.GOV, lists successes of AGOA such as an Ethiopian company, Prospre

International that signed a deal of $70,110 to supply six containers of Niger seed to Bridgeway Trading, a US birdseed importer.

The site cited trade statistics that shows US imports of "edible ice, non-ice cream" products from South Africa "increased from approximately $800,000 in 2003 to $1.8 million in 2004. Because of AGOA, a small African business now has greater earning potential through its access to the US market."

Because of AGOA, it added, Caratex Botswana "reported earnings of more than 6 million dollars in 2003 and anticipates they could reach at least 10 to 14 million dollars. As a consequence, Caratex has grown from employing 500 to around 1,300 people. With the launch of new business attire and jeans lines, Caratex anticipates that it will employ as many as 2,600 people."

Yet, many AGOA critics say the initiative does more for the US than for Africa, which is why it has not been able to make a significant difference in Africa's economic milieu. According to one source, "the acceptance of AGOA means that African countries must renounce not only their economic sovereignty, but also their political sovereignty and means total capitulation to the USA as, amongst other conditions, these countries must pledge not to 'undermine United States national security or foreign policy interests'."

AGOA prerequisites were not a product of negotiations between the US and the African countries but unilateral imposition on the African countries, critics argue. They also contend that insisting on removal of the very few state subsidies in Africa seriously affects the volume of production and the ability of African producers to compete with their US counterparts who enjoy subsidies, grants, loans and tax incentives.

I have been to conferences and workshops on fostering better US policy toward Africa and most Africa advocates are less impressed with AGOA. They feel the US can do better to establish trade with Africa on an equal footing without handing out preferences as to what Africa must do.

When AGOA came into effect, it received a sharp response from T.A. Mushita, an African analyst, who wrote an analysis published in 2001 in the South African Economist describing the US legislation as the "American Growth and Opportunity Act."

Mushita revealed in the article that AGOA was in response to a Cotonou (Benin) agreement between the European Union (EU), African, Caribbean and Pacific nations.

I agree with the position Mushita took on AGOA which has its most deficient clauses in section 104. In this section, a country seeking AGOA eligibility must:

Establish or be making continual progress toward establishing a "market-based economy."

Enact legislation to protect private property.

Incorporate an open-rules based trading system; and

Minimize government interference through measures such as price controls and subsidies and government ownership of economic assets.

"These provisions are not favorable to the interests of the majority of African countries. This is a result of experiences most countries went through within the context of trade liberalization, which involved opening up their markets to global corporations, privatization of national institutes and reduction of government spending. These measures resulted in many governments foregoing programs that provided for social safety nets or cutting basic social services and turning domestic food production to export-oriented cash cropping," Mushita wrote in his article.

This Mushita critique captures the bane of the international struggle for economic parity for Africa which indeed has the resources to do business with the US with the same velocity ratio it is doing business with nations like China. But first, there has to be an acceptance that Africa is a respected trade leader because of its natural and human resources, a worthy cause the Congressional Black Caucus should champion. If certain provisions of AGOA will restrict Africa's ability for speedy economic growth, CBC members cannot be silent.

If CBC sees AGOA as an opportunity to grant Black Americans the avenue to conduct serious business that aids not only their businesses but also indigenous African businesses, its members should fight tooth and nail to ensure AGOA increases instead of decreases that potential.

Another very important organization in this regard is the National Black Chamber of Commerce (NBCC) which boasts 100,000 Black owned businesses in America who account for $100 billion in annual sales.

I am yet to come across raw statistics that indicate how much of those $100 billion are coming from trade with Africa but as NBCC President Harry C. Alford underscored in a Feb. 14 2001 article about the essence of doing business in Africa:

"Entrepreneurship has strong roots in Africa. Thousands of years ago, the trading area that includes the present day nations of Zimbabwe, Congo, Uganda, Burundi, Rwanda, Tanzania, Ethiopia, Somalia and Kenya supplied the world with essential raw materials such as brass, copper and gold. Exotic animals, foods, iron tools and fabrics were also exported to civilizations throughout Asia. African Americans have carried on this tradition of invention and ingenuity. African American business districts in communities ranging from Tulsa and Birmingham to Durham and Harlem were thriving by the turn of the 20th century," Alford wrote.

What happened to Black America since the days of Tulsa Oklahoma, a vivid example of a Black city that thrived economically and sustained itself, remains a valid discussion for Blacks everywhere.

But all hope is not lost. Cities like Detroit are positioned to chart the way for the needed new kind of thinking and pragmatism. One of the most Afrocentric cities in America and the largest bastion of Black electoral power, Detroit and its residents of 900,000 can make a difference in energizing economic cooperation between Black Africa and Black America. The current mayor of the city, Kwame Kilpatrick, indicated in an interview I had with him in 2006 his strong desire to connect with Africa after visiting the continent a number of times.

What is needed is strong political will backed by an informed community that understands that in this era of globalization, Africa stands as the next bastion of economic power.

Therefore, every major Black city in America should have an Africa office not only to facilitate real extensive business between Black America and Black Africa but also to connect on a more intimate cultural level. An Africa office in Detroit, for instance, will serve as a liaison between the city and the continent and help to harness the potentials and resources of African professionals in the city who desire to help move Detroit forward.

In the 21st century, our thinking as Blacks should be as agents of Black transformation who feel the pulse of the prevailing national and international socio-economic dispensation. We have to get to work, inspired by the dictum that if we seek first the economic kingdom, all other things will be added unto us in the Black world.

As I have often posited, a happy and prosperous people are not a product of mere imagination but of a vibrant economic program that creates meaningful jobs and real sustainable wealth for the people.

Chapter Five
The human cost of American chocolates

Despite the menace of stubborn ailments like diabetes, chocolates are widely consumed around the world, especially in the Northern Hemisphere where many – young and old - eat chocolates as daytime snacks or as desserts, crowning their sumptuous meals.

Lovers caught up in the Valentine's Day frenzy exchange roses and, of course, chocolates, and corporations like Wal-Mart and Costco make windfall profits selling these items that have come to be identified with Valentine's Day.

Be it known that I have nothing against eating or exchanging chocolate gifts on Valentine's Day, and I do not begrudge Black businesses that make normal profits on such an occasion. This is because more Black businesses making normal profits could mean creating new jobs, offering better work benefits, paying relevant taxes, ensuring more stable families where basic needs are provided, and better compassionate Black communities providing needed services for seniors and the disabled.

However, I raise questions about the often untouched or ignored issues relating to how some of these gifts are produced.

While Valentine's Day enthusiasts are busy trying to determine if Godiva, World's Finest, Whitman, Russell Stover or Sander's has the best chocolate, Black children in Africa are paying a high price, sometimes with their lives.

Thousands of innocent Black children in Africa are lured into indentured servitude in the coffee cum cocoa industries of Africa - which produce chocolate - with promises of a good living condition and a better future.

For instance, in Ivory Coast, children from disadvantaged homes and economically depressed West African countries are forced into child slavery on the cocoa farms of Ivory Coast.

Statistics have shown that half of the world's cocoa is grown in Ivory Coast with about 600,000 plantations that are exploiting children as young as age nine. In 2000 alone, the US State Department estimated that about 15,000 children were actively enslaved on cocoa farms.

On the plantations, the children are hardly rewarded for their work, and if they dare to protest, they often face severe beatings, chaining and sometimes death.

Some Black Americans were appalled by what the "Blood Diamond" movie illuminated about the plight of Sierra Leonean children in diamond-driven war situation, but innocent children are still victims of cruel injustice in other parts of the Black world and many in Black America hear little or nothing about it. Even those who hear about it tend to act as if they don't care just because such injustice is committed in "distant" Black Africa.

Wondering why the plight of these enslaved Black African children is religiously shielded from public glare? There are people who feed off such abominable travesty and tragedy. Their wealth apparently comes from helping the big companies that import cocoa from countries like Ivory Coast to continue to make lots of money with minimal threat to their profit margins.

In fact, when the Philadelphia Inquirer published a series in 2001 about child slavery on the cocoa farms in Ivory Coast, the US Chocolate Manufacturers Association (CMA) basically said it had no knowledge of the situation.

Their first reaction to the story was typical of corporations that are in business only to make money, with no regard to morality and humanity. They claimed they were not responsible for the modern day enslavement of children in West Africa because they (the chocolate companies) do not own the plantations. They just buy the beans.

So, it wasn't surprising that the cocoa industry used its powerful lobbying machine, engaging the services of former Republican and Democratic majority leaders Bob Dole and George Mitchell, to plead their innocence before Congress.

As usual, a compromise plan was passed to try and force the industry to make some changes in dealing with the cocoa industry in West Africa. I have not seen an evaluation since that Congressional action to determine if the chocolate industry is abiding by the process Congress created. The chocolate industry must abide by the standards of international labor to eradicate child exploitation and slavery in Ivory Coast.

In an essay titled, "2007: A Sankofa Year," Emira Woods, co-director of the Foreign Policy In Focus (FPIF) program at the Washington DC based Institute for Policy Studies, wrote that modern day slavery takes many forms.

"In Liberia, the Bridgestone/Firestone Corporation continues to profit from slave-like conditions in their rubber plantation. Firestone's operations force children as young as 11 years old to work in the fields from before the sun rises to the late day. Used as beasts of burden, these kids typically carry two 75 pound buckets of rubber for up to two miles to storage or collection tanks. Should the children refuse to work, their parents risk losing the measly $3.19 daily wage, all while

Bridgestone/Firestone announces record level profits for 2005 and the first half of 2006."

In 2004 alone, the worldwide net sales for Firestone were said to be in excess of $23 billion, and $9.8 billion of those dollars came from the Americas.

Woods is among passionate activists leading the "Stop Firestone Campaign" against child exploitation in Liberia.

In November of 2005, the International Labor Rights Fund in Washington DC filed an Alien Tort Claims Act (ATCA) against Firestone in US District Court in California alleging "forced labor, the modern equivalent of slavery" on the Firestone plantation in Harbel, Liberia.

ATCA was enacted 217 years ago, soon after the founding of the US. Though meant to deal with human rights abuses outside of the US, ATCA is little known today because only few foreign victims of human rights abuses seek justice in US courts.

"The plantation workers allege, among other things, that they remain trapped by poverty and coercion on a frozen-in-time plantation operated by Firestone in a manner identical to how the plantation was operated when it was first opened by Firestone in 1926," the suit stated.

The million-acre plantation was set up in 1926 when Harvey Firestone secured a 99-year lease on the land in exchange for a $5 million loan to help the government of Liberia repay debt it owed to the US.

Now, 12 Liberian workers and their 23 children, who are anonymous for the protection of their identity, are plaintiffs in the Firestone case.

In an April 26 2007 interview on the Charlie Rose show on PBS, Princeton University economist, Alan Blinder, said the US has not signed on to any international labor standards that bar child labor.
"We tell everyone to," Blinder said, noting he'd like to know why the US urges every other country to sign international treaties but would not append its signature to international labor standards.

There is no need to figure out why, Mr. Blinder. That is the double standards of our foreign policy on display, a practice that has arguably helped America in maintaining its hegemony over global politics and economy.

We give a blind eye to corporations exploiting the resources of poor regions in Third World countries like Sudan, Liberia and Nigeria just as we celebrate greedy corporate bosses who place more value on their profit margins than on the sanctity of human life.

Where is the voice of Black American activism on these serious issues tormenting the Black world? It is ironic that during the month of Black History in February, we in Black America romanticize about Africa but hardly join the continent in its fight against imperial forces that continue to exploit its resources at the expense of majority of its 740 million people who are mostly Black.

Thus, Black Americans have a very important role to play in our march towards the socio-economic transformation of the Black world.

I eagerly look to see Black leaders in America devote some of their activist energy to fighting child slavery in Africa, especially when there is a direct connection between profit worshipping members of the comprador class here in America and their collaborators on the African continent who facilitate the grievous exploitation of natural and human resources in nations like Liberia and the Ivory Coast.

Black leaders and the rest of Black America should take action now so that when history renders its verdict on the true accomplices of modern-day slavery, Black America will be vindicated.

Besides, when Black America takes the requisite redemptive action of speaking out against the exploitation of poor Black children in Africa, those among us who passionately celebrate Valentine's Day will not feel guilty when their partners offer them the chocolate that is often associated with Valentine's Day.

There is a great need for more activists like Emira Woods, and you can be one of them.

How? You asked. By emailing or calling your elected officials, organizing public awareness campaigns, and writing a letter to Bridgestone/Firestone President, Daniel Adomitis, in Ohio to stop the child labor and the environmental destruction of the African country of Liberia.

You can also mail Firestone at Daniel Adomitis, President, Firestone Natural Rubber, 381 W. Wilbeth Rd. Akron, Ohio 44301 or visit www.stopfirestone.org to join the campaign.

Chapter Six

Enough of the natural resource-driven wars

To many among us, presenting a diamond ring to someone demonstrates the pinnacle of one's emotional disposition and social sophistication. It is something to brag about - the price tag, the brand name, the shop where it was purchased, the ambience of its presentation and the attending gestures. Diamonds are greatly cherished but many of them are banners of shame to humanity.

I am not talking about how much it costs to present the diamond to one's partner. If you can afford it – without running into debt - and believe it would please your partner, give her diamond for goodness sake. But while you do so, take a moment to think of the tens of thousands of lives that may have been lost to get that diamond over to us here in Black America. Think about the many children who have been kidnapped and brainwashed into becoming child soldiers in conflict zones of Black Africa, protecting diamond-mining areas for Western corporations to exploit.

By now, you probably have seen the movie, "Blood Diamond," starring Hollywood actor Leonardo DiCaprio. A heartbreaking account of the exploitation of Black Africa's vast natural resources, the film presents its viewers with a glimpse of how some greedy corporations in Western nations are exploiting mineral resources in conflict zones in Black Africa.

Centered around the 14-year civil war of the Black African nation of Sierra Leone, the film ended without justice rendered to the numerous victims of blood diamond exploitation, typical of the battle between the powerless and the powerful. The corporations that fan the embers of conflicts in Black Africa by providing arms to rebel groups in exchange for mineral resources are still free of prosecution, even when ample evidence point to their connection to not only extracting blood diamonds, but also selling arms to rebel groups.

However, it is important to note that Sierra Leone's war over conflict diamonds goes back to 1991 when hundreds of mercenaries crossed the Liberian border and attacked villages and towns in the southern and eastern parts of the country.

Then in 1992, the Revolutionary United Front (RUF) led by rebel leader Foday Sankoh, a student activist in the 1970s who was imprisoned and later exiled in Libya where he was mentored by Libyan ruler Muamar Ghaddafi, seized Kono, the diamond mining capital of Sierra Leone. The country's National Provisional Ruling Council initiated "Operation Genesis" to flush out the RUF rebels in what became a protracted battle.

In the words of Greg Campbell, "The RUF began its jewelry heist in 1991, using the support of neighboring Liberia to capture Sierra Leone's vast wealth of diamond mines. Since then, the rebels have carried out one of the most brutal military campaigns in recent history, to enrich themselves as well as the genteel captains of the diamond industry living far removed from the killing fields. The international diamond industry's trading centers in Europe funded this horror by buying up to $125 million worth of diamonds a year from the RUF, according to UN estimates."

But did people care enough to look beyond the surface? Was the conscience of the international community pricked? Did diamond consumers in Detroit, Chicago, Atlanta, Washington DC and St Louis hold the diamond industry accountable? No, of course!

"Few cared where the gems originated, or calculated the cost in lives lost rather than carats gained. The RUF used its profits to open foreign bank accounts for rebel leaders and to finance a complicated network of gunrunners who kept the rebels well-equipped with the modern military hardware they used to control Sierra Leone's diamonds. The weapons - and the gems the rebels sold unimpeded to terrorist and corporate trader alike -allowed the RUF to fight off government soldiers, hired mercenaries, peacekeepers from a regional West African reaction force, British paratroopers, and until recently, the most expansive and expensive peacekeeping mission the UN has ever deployed," wrote Greg Campbell, author of "Blood Diamonds: Tracing the Deadly Path of the World's Most Precious Stones."

While evidence suggests that this war was despicably funded by the international diamond industry, there was no international intervention until 2001 when the UN came in. The United States stayed out of the scene while thousands of Black Africans died and children maimed and amputated with machetes and axes by the RUF which was being funded by the international diamond industry.

At the early stage of the war, the international mainstream media was nowhere to be found. The Black American media was nowhere to be found. Instead, we saw after the war had ended the birth of the Kimberly Process, "an international certification scheme that regulates the trade in rough diamonds. Its aim is to prevent the trade in conflict diamonds, while helping to protect the legitimate trade in rough diamonds."

With 45 global participants, including the US, the Kimberly Process came out of a May 2000 meeting in Kimberly, South Africa, to address the mindless mining of natural resources in conflict areas.

Today, Kimberly is struggling to hold accountable those behind the buying of blood diamonds, the major cause of conflicts in the

African countries of Angola, the Democratic Republic of the Congo, Liberia and Sierra Leone.

All of these countries are resource-rich with diamonds that are easy to mine and smuggle. These nations are among the richest on the continent, but have mostly been known for turmoil for several decades.

Liberia which has historical links with Black America was allowed to go into near extinction in part because of its natural resources.

Blood diamonds may be a new form of contemporary African exploitation, but the continent has long been raped of its resources, dating back to the Berlin Conference of 1884-85 when Africa was divided among Western powers. The Congo's vast mineral resources were major driving force in the partitioning of Africa in which Belgium and its King Leopold played a major role.

Since then, coupled with the assassination of Congo's Patrice Lumumba - a leading Pan-Africanist - the Congo has known no peace until recently. Endowed with coltan, an essential mineral for making cell phones, computer chips and nuclear reactors, the Congo has long been the theatre of Africa's longest running mineral resource-driven warfare.

Africa produces about 65 percent of the world's diamonds. That is $8.4billion a year that goes to various economies of the world. But the other side of blood diamonds, which has not always been told, is the complicity of the US media, which has shown no interest in Black African affairs, especially since its coverage of issues and events has often been predicated on the policy thrust of Washington. The media will always follow the policy direction of Washington. Thus, as the US seemingly has no strategic interest in Africa, the media lets major humanitarian disasters and natural resource-driven wars slip under the rug even if thousands are being killed daily.

In Rwanda, where former US President Bill Clinton admitted "dropping the ball" and letting 900,000 Black Africans get killed with no US intervention, the US media was unrepentantly absent. And like "Blood Diamond," the Rwandan human carnage came to our living rooms through the movie "Hotel Rwanda" after 11 years of the genocide.

While completing a 2003 foreign policy fellowship in Washington DC, I realized that most reporters and editors of major newspapers in the US were hardly interested in covering workshops and lectures on critical Third World issues.

I recalled attending a reception at the Robert F. Kennedy Memorial Center for Human Rights held for former Liberian Catholic Archbishop Michael Francis. He was in town to brief Congress about the dire situation in Liberia at the time, and to persuade

Congress to send troops to Liberia to quell the rising rebel conflict.

In a solemn but firm tone, Francis charged that on the day of reckoning, the Lord will ask if America's good works around the world are only driven by "strategic interests."

There were very few media outlets at the event.

Black genocide

In 2000, an exhibit called "Genocide Warning: Sudan" was unveiled at the Holocaust Museum in Washington DC. Jeff Drumtra, the then senior policy analyst on Africa for the US Committee for Refugees, chided the mainstream media in America for failing Sudan.

"Journalists cover events that are extraordinary. Events that are unprecedented. Sudan is full of records - awful, grisly records. Sudan is full of headlines - grim headlines. Yet American journalism largely ignores Sudan. In what way is Sudan not worthy of coverage?" Drumtra painfully asked reporters at the museum.

Drumtra told the journalists that during the year 2000 alone, the Arab controlled Sudanese government dropped bombs 115 times on towns and villages.

The Sudan genocide exhibit was meant to educate the American public. Because Sudan means the land of Black people, Drumtra threw a challenge to the Black American press about its reticence on the crisis. "Where is the Afro-American press?" he asked. That is the haunting question for today's Black American media as far as African issues are concerned.

Today, the Sudanese government-sponsored Janjaweed militias are carrying out genocidal acts in the Darfur region of Sudan against Blacks Sudanese. Up to date, no serious and situation-impacting punitive measure has been taken against the Sudanese government since former Secretary of State, Colin Powell, finally called what was happening in Darfur "genocide."

Black American leaders have been basically silent on the Darfur region. The genocide has been treated as a remote conflict. It has yet to garner the kind of political activism the fight against apartheid in South Africa received. Sometimes, I think it reflects our reluctance not to confront problems that do not have visible White versus Black racial underpinnings. When it was South Africa, we created a movement. Now that the people of Darfur need Black political leaders to stand up, most of them are nowhere to be seen. Black American leaders need to show some muscle to pressure Washington for some serious, well thought-out intervention on the continent.

And Black Americans can send a strong message to the Sudanese government in Khartoum that the onslaught against Black Sudanese in that nation must stop. One way to send that message is to start exercising what a Michigan State Senator, Hansen Clarke, is exploring: Divestment.

Clarke, an African American, wants to urge the State of Michigan not to do business with any corporation that is engaged in direct or indirect business dealings with Sudan. He is behind a bill to amend 1965 PA 314 entitled "Public employee retirement system act" to require all the Michigan pension funds to be divested from any company doing business with the Sudanese government.

Clarke is taking leadership on an issue that many are shunning. As history is being written, a lot of people who should have done something would find themselves guilty of the blood of thousands that have died in Sudan.

"Throughout history, it has been the inaction of those who could have acted, the indifference of those who should have known better and the silence of the voice of justice when it mattered most that has made it possible for evil to triumph," according to former Ethiopian leader, Haile Selassie.

After-word

Bankole Thompson is a pragmatic journalist who lives the mantra of Freedom's Journal, the first Black newspaper founded in New York, which read, "We wish to plead our own cause."

In his second book, *A Matter of Black Transformation*, Thompson offers a bold-faced approach to issues facing the Black world.

Throughout my years in international diplomacy and serving as a legal counsel to African political institutions, I have traveled all over the world and seen the despicable socio-economic carnage our people are caught up in. And my work around the world has been to address the crisis in education, health, unemployment, economic disparities in the Black world, conflicts in Africa and other germane issues facing us as a people.

In *A Matter of Black Transformation*, Thompson, a journalist whose work I greatly admire, issues a clarion call to everyone engaged in the Black struggle to chart a new way of thinking in solving our many problems. The engaging book and its frantic opening chapter on "Pushing frontiers of Black education" is at times biting, but a downright truth telling style Thompson is known for in his journalistic work.

He makes a strong case for Black America to connect with Africa, a discussion that more often than not has been relegated to the back burner. Africa's progress is central to the development of Black America.

Dr. Martin Luther King Jr. once said that the "liberation struggle in Africa is tied to the struggle for Jim Crow in America." That underscores how King viewed the fight for civil rights and economic justice for Blacks in America. Black transformation is a global struggle, as King understood it.

When Ghana gained independence on March 6, 1957, Dr. King, and A. Philip Randolph attended the historic freedom hailing ceremony of the first country in West Africa to break from European colonialism.

In a front-page editorial, the Pittsburgh Courier trumpeted Ghana's independence saying, "When we, American Negroes, shake hands with Ghana today, we say not only 'Welcome!' but also, 'Your opportunity to prove yourself is our opportunity to prove ourselves.'"

Thus, author Thompson in his book detailed how Blacks in America can be working closely with a continent so rich and endowed with natural resources. He expounds on the notion of globalization and its impact on Blacks on every page.

Thompson explains as you go through the pages of his book that Black American businesses should be actively engaged in Africa's economic development through partnering with their counterparts in Africa, and making the most of the Africa Growth and Opportunity Act. He cited White corporations who are doing business on the continent to underscore the fact that Africa offers attractive investment opportunities.

Even China, the book reveals, has boldly taken its seat in harnessing Africa's vast natural resources and trade opportunities. In 2006 alone, as the book shows, China's trade with Africa amounted to about $50 billion.

Thompson therefore throws a challenge to Black America to step up especially when we command an annual buying power of about $700 billion.

But first, he offers in the book a requirement that both Africa and Black America should engage in a process of meaningful education to understand the intricacies of each other's socio-political and economic realities, needs and challenges and the opportunities presented by such understanding within the context of globalization.

Black American businesses will be convinced that Africa is a good spot for business investment when they gain the empowering and emboldening understanding of the difference between the "unattractive" Africa of the media and the Africa that boasts resplendent ideas, endowments, opportunities and achievements.

Thompson urges Washington Black lawmakers to do more to enhance provisions of the Africa Growth and Opportunity Act. He also challenges us to use the same energy we brought to the campaign against the South African apartheid, in helping to bring closure to the human carnage in Darfur.

Thompson proceeds to raise awareness and consciousness on other issues in Black America like the Black family, maintaining that they need to be tackled pragmatically and futuristically.

A Matter Of Black Transformation is required for anyone who wants to understand 21st century problems facing Black people around the world and how to solve them. With intellectual dexterity and knowledge of the international economic and political system, this respected newspaper editor in Black America has made a case for a real and rewarding transformation of our communities.

Bankole Thompson has clearly built on the foundation laid by my brother Alex in *Roots* and I am happy he is on a mission to help strengthen the socio-economic ties between Black America and Black Africa in ways that will make the Black world glow.

George W.B. Haley

Former US Ambassador to The Gambia
Appendix i

Interview with Dr. Jeffrey Johnson, Black family expert

Dr. Jeffery Johnson is the foremost expert on the Black family, especially the issue of fatherlessness. His Washington DC-based organization, National Partnership for Community Leadership (NPCL), is a leader in providing resource and training for Black men who risk being hurled away to jail for not caring for their children.

Johnson talks about the state of Black fatherhood, an important ingredient in building strong Black families and hence strong and better Black communities.

Bankole Thompson: What is the state of Black fatherhood in America today?

Jeffery Johnson: The state of Black fatherhood in America today is one of grave concern. Most significant is the absence of fathers in the lives of children and youth. By their 16th birthdays, 80 percent of Black children will live in households without their fathers and 36 percent of White children will do so. Additionally, female-headed households have poverty rates fives times those of two-parent households, and the probability of incarceration by age 14 is 30 to 40 percent lower for children who grow up with both of their biological parents than for children who grow up in any other family arrangement, including a family with a stepfather.

What in your opinion have been the contributing factors to fatherlessness in our communities?

There are a number of contributing factors to the absence of fathers in families and communities. First, for more than a generation, many of our young people have grown up without married parents and fathers in the home. The result has been a lack of commitment to marriage and sustained engagement of fathers in the home. Young parents are essentially modeling what they were exposed to, having children outside of marriage and father absence. To many of them, this appears to be normal behavior. A second factor is the lack of jobs that pay a family sustaining wage, particularly for low-skilled men. For the better part of the 20th century, the American manufacturing industry was the hub of employment for Black men. These jobs paid good wages and offered fringe benefits such as health care. The decline in blue collar employment has disproportionately affected Black men, contributing to their marriage-ability (both an external and internal view) which has had an adverse impact on family formation. While marriage rates among African-Americans have declined over past twenty years, births out of wedlock have not. Seven of every ten Black children born in America are born out of wedlock. The prospect of Black couples marrying once a child is born is very low.

A third factor impacting father absence in the Black families is the high rate of incarceration among young Black men. There are a number of studies that indicate that in excess of one-quarter of all Black men in America between 16-35 years of age are under the control of the criminal justice system (jail, probation, prison). These studies also point out that that there are more Black men in prison, than on college campuses. These high rates incarceration are associated with the sale, distribution, and use of illegal drugs, and mandatory sentencing. The result is that if a father is incarcerated, he is not in the home and therefore is not a visible role model in the home for the children. There are many factors; however, I think the ones I've mentioned are key contributors to fatherlessness.

Where do you place the National Partnership for Community Leadership in this phenomenon of fatherlessness?

NPCL is in the forefront of educating the nation on problem of father absence and its causes. We are also playing a leadership role in building the capacity of family and human service agencies to target and successfully work with men and fathers.

Our national family policy has only recently included men and fathers as a focus of both prevention and intervention efforts. For the most part, every public system response to men is of a punitive nature such as jail or child support. I believe we can teach young men and women how to be responsible adults and parents. However, to do so will require a commitment of resources and creative partnerships among agencies who work with families. NPCL through its capacity building efforts and annual international fatherhood conference is raising awareness and showing people how we can fix the problem of father absence in families and communities.

Why does NPCL see it fit in leading the challenge of strengthening Black fathers?

When NPCL was established in 1996, we started with an understanding that there was a big gap in the number of agencies committed to strengthening families with a focus on reconnecting fathers. Our mission from day one has been working with low-families and "deadbroke dads." The latter term, created by NPCL, denotes a willingness on the part of poor fathers to support their children, even though they are without the financial resources to support them. This is in contrast to "deadbeat dads" who have financial resources but are unwilling to support their children. We believe that by stabilizing the father's role in families, we can increase the well being of our children in health, school, and the community.

What has changed with the Black family today from that of the past?

The biggest change in the black family since the reconstruction period after slavery is the decline of marriage-based families. In 1890, eighty percent of all Black families were husband-wife families. This pattern continued through the early 1960s. Today however, seventy percent of all black children are born out of wedlock. Without a doubt, there has been a dramatic change in Black family structure.

How does the image portrayed about Black men in the mass media contribute to the problem?

There is a saying, "If it bleeds, it leads." I believe this statement reflects how the media reports on Black men. More often than not, Black men are portrayed as angry and dangerous people to be feared by the public. While Black on Black crime is a serious problem, it only represents a small segment of the Black community. Most African-American men and women are law abiding citizens with dreams and aspirations similar to those of most Americans. However, when the media only emphasizes the negative, Black men have historically gotten the shaft. I tell people all the time, I love being a Black male, but it is inconvenient sometimes.

Some men have complained about the child support system. Is the system fair?

I agree that the child support system has been unfair to low-income men. Notwithstanding, my belief is that men and women who owe child support and are able to pay should pay. Children need the financial and emotional resources of both parents. The problem is that the child support system was designed for middle class parents. Unfortunately, neither the child support system nor the courts differentiate between the fathers who have the ability to pay and those that can't. I believe it makes no sense to incarcerate a low-income or no income father for nonpayment of child support when he never had the means to pay in the first place.

What has your organization achieved in helping families combat fatherlessness?

NPCL has done quite a bit since 1996. We managed for several years the Ford Foundation's Strengthening Fragile Families Initiative. This project engaged prominent researchers and practitioners in efforts to come up with solutions to strengthen family functioning among low-income never-married parents. Based in part on the Ford Foundation project, NPCL played a significant role in passage of the first national fatherhood legislation. For the first time in our nation's history, there is $50 million dollars available annually to family and human service agencies to target and provide services to fathers. We also for the last nine years have sponsored an annual fatherhood conference. This event attracts people from around the world who are interested in strengthening families. Finally, we train about 1000 agency staff per year on how to work with fathers. Taken together, we believe NPCL is making a difference.

Where do you see the Black family heading to?

One of my favorite scriptures is, "Where there is no vision, the people perish." I am a visionary and I believe we can rebuild strong families. If we could survive slavery, I know we can overcome any challenge facing our community today. With all the brain power and resources we have at our disposal today, we can tackle any obstacle that confronts us.

What is the biggest motivation for you tackling fatherlessness in the country?

I am a former school teacher. My main motivation in working on father and family issues is love for children. When children grow up in loving families, they are safe, secure, and happy. This allows them the best chance of reaching their highest human potential. The love of children is what motivates me.

Appendix ii
Interview with Adrianne George, Black American entrepreneur in Europe

Adrianne George, an African American who migrated to Europe is making waves building businesses and connecting with Blacks in the Diaspora. Adrianne talks about her experience as an educated Black woman living in Sweden, and the complexities of life on a continent different from America.

Bankole Thompson: What drew you to Europe?

Adrianne George: My first trip to Europe was to London at the invitation of Mad Professor. I met him while I was working at RAS Records. We distributed his music in the US. He's a music producer in London and, at least at the time, was the owner of the largest Black owned studio.

What has been the reception since your arrival in Sweden?

In Sweden, like everywhere else I've been in Europe, I'm an American and thus received as such.

The reception in Sweden has been warmer than the reception I received in Belgium in 2002 on the eve of the war with Iraq. People were downright hostile and very anti-Bush.

You are involved in connecting with Blacks, especially women in Europe who are influencing and making changes in society. Why?

I've met several Black women entrepreneurs living in the UK, France, the Netherlands and Germany. They are a combination of American expatriates and Afro-Europeans born and raised in Europe. We all talk about how good it is to connect with each other because at the beginning and end of the day, we're Black women and we need to support each other. For the successful expatriate women I've met, being a foreigner means easily surpassing the low expectations placed on immigrants, not necessarily Black women. Politically, there are Black women in government in the UK, Belgium and Sweden, and maybe other countries too. There may be one in France soon as well. They inspire me as do all the other Black women who succeed in predominantly male and White arenas, simply because no one expects them to and I respect their struggle and individualism. I don't have to agree with their politics.

In America, we joke sometimes about what it means to be Black. Can you say the same thing in Sweden?

I was surprised to see so many Black people in Sweden. A lot are first generation immigrants while others have been in Sweden since their childhood or were born in Sweden. I've been told that Black people are thought of as hard working in Sweden. But I haven't met enough Black women born or raised in Sweden to know how they define what it means to be Black. I do know that my family subscribed to *Essence, O* and *American Legacy* for me because I haven't seen any publications for Black women in Sweden. I'd love to know of some! You can get a magazine about Africa here though. I'm embarrassed that right now, the name escapes me even though I read several issues when I lived in Brussels. I think it is *The New African*.

How are Black women in Europe in your opinion faring?

Not as good as I would like, but not as bad as one may think. However, in a lot of large cities, the residents from former colonies are still alienated from mainstream society just as many Black Americans feel alienated from mainstream society in the States. You probably heard about the German soldier who trained his men to think of scary African American men in order to gear themselves up to shoot to kill. He was fired quicker than Imus, but I imagine that attitude must spill over to Blacks, including Black women that live in Germany as well. I wish more Black women in Europe took advantage of the social system and earned university degrees. I am sure many do, but I have met so many who don't. Within the last 3 or 4 years, French television hired a Black female news broadcaster for the first time. By the way, they hired their first male broadcaster around the same time. But I honestly think that Black women are accepted in Europe than by White America.

So, is racism as blatant in Europe as in America?

I would probably be naive if I said no, and I am sure there are lots of horror stories, but in my case, I haven't experienced any. But perhaps, that isn't true because whenever I am stopped (in a language, I don't understand) at an airport in Europe after I have gone through customs and passport control, as soon as the official hears my American accent, I am waved through. I always ask: what did you want? They say everything is fine. I think you have to tackle the racism question country by country. For instance, Malta asked the EU for help in dealing with the large number of African immigrants who come in boats. They feel overwhelmed, but Italy and Spain complain as well. Those situations remind me of the Haitian boat people. In the face of such desperation, where is the humanity? France has problems with West and North African immigrants. Remember the riots in the suburbs not too long ago after the police chased two youths to their deaths? A lot of the French hip hop music I've heard sounds angry. My Swedish isn't good enough yet to understand Timbuktu's lyrics.

Black women are having a nice time then in Europe

I think I covered this in the previous question. But I can add that it can be hard to find make up for Black women even in cities with vibrant Black populations. I researched this while I was getting a master's degree in business from Boston University in Brussels. So it's like the beauty companies are ignoring what would be a lucrative market. The US finally realized that about 15 years ago. I consider that racist, to ignore a market segment because you don't think they are important.

What is the opinion about African Americans in general in Europe?

I am always amazed at the depth of knowledge a lot of Europeans have about our music, from gospel, especially blues and jazz, to soul, R&B and hip hop. The fact that they love and respect our music makes me proud. There is a group for Black women called the Afro European Sisters Network and Sandra, the Afro-European sister in the Netherlands who runs it, says she wants to inspire Afro European women to excel and succeed like Black American women have and do. Condi Rice is held in high regard even though Bush isn't. I may be overly generous, or maybe I've been lucky, but I haven't met anyone who has said anything bad about African Americans or treated me with disrespect. I can say that the press only seems to cover bad news about Africa though.

Do they think racism is responsible for many of the African American problems?

Europeans know an awful lot about American history as well as their own part in the transatlantic slave trade. They know about the civil rights movement as well, but I am not too sure they know about the current state of affairs. They see Oprah, and Condi, Beyonce and Halle, and think we're doing OK.

Do you intend to come back to America?

I don't know if I'll be an expatriate for life. As it is now, I come home twice a year and as needed by my family. I can say that there is a lot about the European lifestyle that I like. For instance, I like the work-life balance. I also like being able to visit different countries as easily as we visit different states in the US. But I do get terribly homesick at times and frustrated by the lack of personal drive I see in a lot of people. For that reason, I know that the US will always be in me no matter where I live.

Appendix iii

Interview with Tony Mottley, producer American Black Journal, PBS

If you are disappointed with the media, all hope is not lost because outlets like American Black Journal (ABJ), the weekly Sunday television program that airs on WTVS- Detroit PBS station, are making a difference. The show provides a breath of fresh independent journalism, untainted by the mud of political expediency. Guests are interviewed by host Cliff Russell, a veteran journalist. ABJ is believed to be the nation's longest running public affairs show on Blacks.

The show, now in its 37th year, first began in 1968, when it was called "Colored People's Time," serving as a public forum for Blacks at a time of racial turmoil across the nation.

Some of the show's popular hosts included, Ed Gordon, Ron Scott, Tony Brown, Juanita Anderson, Jim Ingram, Darrell Woods and Darrell Dawsey. Former Detroit Mayor Coleman A. Young, South African Archbishop and Nobel Laureate Desmond Tutu, Roots author Alex Haley are among notable guests interviewed on ABJ.

ABJ producer Tony Mottley explains why ABJ is still holding on.

Bankole Thompson: What has kept the journal on air for almost forty years?

Tony Mottley: Thankfully, Detroit Public Television has remained committed to American Black Journal. Moreover, there is a need for the show in our community. The need reaches beyond Detroit. We have viewers from Vancouver, Canada all the way to the Virgin Islands. Black people of all stripes and a segment of the White community respond to issues that affect Blacks when presented in an intelligent and respectful manner.

Why should people watch American Black Journal?

ABJ is a program that addresses the concerns of our community in a serious manner. You should watch because our goal is to present an unfiltered view of life in America from a Black perspective. Health and nutrition, finance, politics, entertainment, sports, leading authors, politicians, journalists and other public figures are all a part of our weekly mix. We not only tell you what happened, we present ideas about how to react to situations and offer a perspective of what action, if any, should follow.

How did you become producer of the journal?

I came to DPTV in 1989 and worked as an intern on Black Journal. It was a local election year and we covered former Mayor Coleman Young's last election campaign. After my internship, I worked as a stringer for ESPN; I produced a radio show on WDET, Detroit's NPR affiliated. I came back to DPTV in 1992 and produced a documentary on youth violence that received an Emmy Award. In 1993, I was hired to produce Black Journal.

What is the biggest motivator for you being a producer for fifteen years now?

I think there is still so much work left to do. Our people still suffer from many of the same maladies they've battled since the program began. In addition, it's so exciting. Being on the frontline of issues and being a witness as news stories break is exhilarating. Our people still own and control the same percentage of wealth we did on the eve of the Civil War. That's a painful statistic. In the face of all the supposed new Black successes, Black people as a group are still lagging behind. I want to be around when that changes. I want to see my team win.

What role do you see the Black Journal playing in Black media?

ABJ is an independent voice of the community. We have to speak truth to power. We are a safety net for our people. When the so-called mainstream media rakes Black folks over the coals, we are there to set the record straight. We are also a safe haven where other Black journalists can come and speak freely on issues.

Do you receive reactions on what you report and discuss on the Journal?

Over my fifteen years as producer, there are two consistent reactions to our program. The guests like Wynton Marsalis or Berry Gordy or Cornel West are shocked to see an all-Black team producing a program. They are all stunned that we've been doing it for so long. Our viewers depend on us for information they can use to improve their lives. They want to know the name of a book and where it's available. They want to discuss subjects that have been presented. Teachers want to get copies of the show to play in their classrooms. They thank us for touching issues that matter to them. They call, they write letters from dorms and from prison. They send email.

What would happen if there was no American Black Journal in the largest Black city in America?

It would be a dark day if there were no Black Journal program. Detroit is the capital of Black America and we are Black America's program. No other show has the legacy or the present day tenacity and commitment to advance the cause of Black folks. Without question, Blacks would be less informed and more misinformed if ABJ was off the air.

What are the challenges of producing the journal?

Our challenges haven't changed. We work with very tight deadlines, with very small budgets and very little acclaim. This is a labor of love. I always tell my interns, there is no money in Public Television. We aren't here for the glamour, glory or cash. Indeed our passion is for the work that we do for our people.

What is your view of the state of the Black media today?

Corporate greed and integration have stymied Black media. There are so many Blacks who are willing to be the next Thomas Sowell or Armstrong Williams. To a lesser degree, there are more Blacks who wish to be the next Tavis Smiley. To me, that's a compromised form of Black Media. Certainly, the Tavis Smileys of the world do good work, but as Dr. Claud Anderson once said, racism is a competitive sport. When your team loses, we all lose. Any one who participates in the marginalization and compromise of the Black condition is helping the other team. The notion that Blacks should carry the torch of multiculturalism and integration is dangerous. Blacks (the Black media) should advance the cause of Black people. We are too willing to embrace this modern false sense of diversity to demonstrate how down we are with everybody while we need to focus on our own issues.

Is the Black media relevant today?

In many regards no. (See above) We need to set our own agenda. I mean an agenda that speaks to moving Black people forward. When we start with the Rainbow or people of color notion, we are losers. How can we be relevant when our goal is to simply get along with and seek acceptance by people who are *whipping our collective asses.* We in the media are too consumed with celebrity. For every George Curry, there are 100 handkerchief head Negroes willing to carry White folks water.

How can we be relevant when we don't address our own issues? We all know Black reporters and producers who are given dictates about focusing too much on "urban" issues. It's true and we all know it. Too often we give in rather than fight.

Appendix iv
Interview with US Representative, Carolyn Cheeks Kilpatrick (D-Detroit), head of the Congressional Black Caucus

There has to be transformation within our institutions and leadership cadre if any light is expected under the tunnel. US Representative Carolyn Cheeks Kilpatrick (D-Detroit) is the head of the Congressional Black Caucus, elected after the sweeping Democratic victory in the 2006 Congressional election. She is poised to effect changes within the Congressional Black Caucus.

Bankole Thompson: What is your vision as CBC chair?

Congresswoman Kilpatrick: As the 20th Chair of the CBC, I want to mobilize America around this year's theme: "Change Course, Confront Crises, Continue the Legacy." Since the inception of the caucus in 1971, its core mission has been to "Change Course" for African Americans and others by working to close, and ultimately eliminate, disparities that exist in education, health care, housing, and other areas. In order to address these issues, we must "Confront Crises" like the war in Iraq to strengthen and maximize the potential of our families and communities. We must also "Continue the Legacy" of the CBC and all those leaders on whose shoulders we stand. Our position has always been, "The CBC has no permanent friends, no permanent enemies; just permanent interests."

What in specific terms do you plan to achieve in your tenure as head of the CBC?

Throughout the duration of the 110th Congress, the CBC will focus on four priorities:

- Empower and Mobilize America's Youth — We will work with young leaders between the ages of 18 and 40 to increase their access to the process and provide them with the resources to set the agenda in their communities.

- Build an International Internet Based Communications Portal —We will establish a national communications network to educate, engage, and empower our community and build stronger relationships with our natural constituencies. We want to be able to communicate with communities all across the country and mobilize them for instant action on issues.

- Continue the CBC Outreach Program — We will continue to travel around the country to talk to America's families about issues such as education, jobs, alternative energy, and prescription drugs that affect their quality of life.

- Impact the Alternative Energy Discussion — We will work with Black farmers to help identify and produce crops that can be used to produce alternative sources of energy, generate revenue, and address issues related to global warming.

What difference does it make for a woman in your capacity to head the CBC?

The 110th Congress is a pivotal time in American history. Women are taking their rightful place in leadership and building bipartisan support for the Democratic agenda. For example, we have the first female Speaker of the House, Rep. Nancy Pelosi. We have the first woman and African American to serve as Clerk of the House of Representatives, Lorraine Miller. The CBC is proud to honor the life and legacy of the late Rep. Juanita Millender-McDonald, the first African American woman to chair a standing committee in the history of the United States Congress. We will continue her work. Last, we celebrate the only African American woman to chair a full committee in the 110th Congress, Rep. Stephanie Tubbs-Jones. Women and men have certain intrinsic characteristics. We tend to promote cooperation and collaboration rather than confrontation. As mothers, we are naturally caretakers and protectors. I think these traits will prove valuable as we fight to take our country in a new direction. It is an honor to serve alongside these great women as we shape public policy and work to preserve the American dream for our children, grandchildren, and future generations.

What do you think have been the shortcomings of the CBC?

The CBC has been viewed through a keyhole instead of a door. People tended to think that we represent only African American interests and called on us when there were "Black" issues. Traditionally, African Americans have been denied access to key leadership positions. This lockout marginalized the exceptional collegiate and professional backgrounds of many Black legislators. However, the pendulum has swung. Many CBC members are in positions of power and setting the agenda for several of the nation's top priorities. Our 42 members include the Majority Whip, four full chairpersons of standing committees, and 17 subcommittee chairs. We represent more than 40 million people of diverse racial and ethnic backgrounds in 26 states. Ten of our members represent districts in which African Americans are not the majority. We will leverage our political capital to hold America to the promise it has made to all of its families.

The CBC was working on a House sub-committee on Katrina. What is the status on that?

To date, the House has passed 10 bills focusing on Hurricane Katrina and Rita recovery. The CBC's charge is to enable the return home efforts of evacuees through federal initiatives. Specifically, we want to rebuild community infrastructure, focusing on facilities and programs for youth, seniors, and people with disabilities (e.g., child care, K-12 public schools, colleges & universities, healthcare and hospital networks). Our government must revamp and reform its level of preparedness and responsiveness to avoid future debacles through which America's families suffer from the loss of life and livelihoods.

What do you intend to leave as legacy for heading the CBC?

I plan to use my two-year term to construct a blueprint that will afford administrations to come a systemic outreach strategy and effective communication plan to deliver key messages. By strengthening our pipelines to leadership on and off Capitol Hill, the Congressional Black Caucus will continue to foster relationships and serve as a touchstone for generations to come.

Appendix v
Interview with Kenneth Harris, founder International Detroit Black Expo

The International Detroit Black Expo led by Kenneth Harris, a graduate of Clarke Atlanta University, is connecting local businesses not only at the local level but also at the global front.

A recipient of the US Small Business Administration's Michigan Business Champion of the Year Award, Harris is injecting a new way of thinking and doing business into Detroit's business milieu. He organized a successful Buy Black Weekend campaign during memorial weekend May 25-28, 2007 at the Detroit Cobo Convention Center that assembled one thousand businesses to meet Black customers.

The result was 50,000 Blacks who answered the call to support businesses of their own kindred. It was an eye opener to some who had no idea how many Black businesses operate in Detroit and the kinds of services and products they offer.

However, like Louis James, Ken Harris believes the true success of Black businesses in America is tied to their connection and collaboration with African businesses.

Here are Harris' thoughts on key issues relating to Black economic empowerment across the globe.

Bankole Thompson: Is Black economic empowerment achievable?

Kenneth Harris: Yes, but it must first begin with creating institutions that can help mobilize, educate and provide resources for black-owned businesses. But, more importantly, these institutions must become conscious of a unified Black economic agenda and strategy towards economic empowerment in the African American community.

Where is the Black struggle at present?

Access to capital is where the struggle exists in Black America. The limited resources available towards African American businesses and start-up firms co-exist with a very low retention rate of sustainability with black-owned businesses.

There has always been a debate on whether civil rights take precedence over economic rights. Where do you throw your allegiance?

Civil rights and economic rights coexist in my opinion. You cannot have civil rights without economic rights and you cannot have economic justice without civil rights. Civil rights allow access to resources, capital and expansion, while economic justice provides the means to produce, manufacture, trade and distribute products and services without global limitations. Economic justice helps businesses become part of the true American dream in a purely democratic, capitalistic foundation set by the United States.

Economic empowerment has not been given the kind of platform civil rights enjoys today. Why?

Economic empowerment will never be given the kind of platform civil rights enjoy today until we are able to maximize the 678 billion dollars of black buying power in America. Until we are able to consume less and produce more, we will never be able to circulate the Black dollar effectively in the Black community.

Why must Black people be concerned about economic empowerment if we can work and pay the bills?

Black folks have always been entrepreneurs since the time of slavery, but since integration, we have allowed the corporate complex to demean creativity and independence. Corporate experience is an excellent tool to perfect your craft and become an expert in your chosen field of endeavor.

Once the knowledge and expertise have been perfected, it is only essential that entrepreneurs form businesses that can services the wants and needs of the African American community as an economic empowerment agent.

When we create a culture of entrepreneurship in the African American community, we can then move towards the creation of generational wealth and revitalization of our neighborhoods, schools, churches and institutions.

Where do you place Africa in this discussion of Black economic empowerment?

Africa is the most fertile land in the world with a vast amount of natural resources. I truly believe that although we were slaves to a nation for more than 400 years, we were brought here to the US as the lost tribe who will make their way back home to rebuild, reclaim and develop Africa as the original paradise. When we are able to make a distinct trade connection between Africa and the United States, we African Americans can then begin to talk about divine economic empowerment through a distinct strategy and bridge between the two parties.

Do Black Americans see Africa as a strategic place to do business?

I think the lack of knowledge and images portrayed by the mainstream media suggest to Black Americans that Africa is the wrong place to do business. But Africa is the exact opposite. African Americans who have visited Africa and built relationship with Africa find it empowering from a sense of culture and economics to invest in

Africa. Again, once African Americans create a distinct institution that will forge a profound economic relationship, more Black folks in America will engage Africa as their first option to do business.

What makes the International Detroit Black Expo different from all other previous organizations that have existed in the past?

The International Detroit Black Expo, Inc., is progressive in the sense that it operates from the bottom-up, grassroots perspective interwoven with a technological sophistication and innovative strategy. IDBE is the first organization of its kind to clearly embrace the concepts of economic empowerment as its mission and vision, shifting the pendulum from civil rights to economic rights. But, IDBE also eliminates the bureaucratic way of doing business and connects African American entrepreneurs directly to the source of empowerment needed to expand, capacity build and provide access to capital, while giving access to the global marketplace. IDBE has become the centerpiece for economic and entrepreneurial thought, strategy and development.

Are you reinventing the wheel?

The International Detroit Black Expo, Inc., didn't re-invent the wheel, we just added to the concept; real economic empowerment concepts that affect people intimately and help to create a new generation and culture of entrepreneurially-minded Black folks.

Why does Black economic empowerment matter to you?

Because now is the time for Black folks to claim ourselves as true Americans, while uplifting ourselves as true global players. If we are going to create generational wealth and sustainable global businesses and institutions, we must embrace economic empowerment and entrepreneurship as a first option towards that dream.

I believe that economic empowerment will free us from economic slavery and provide a domain for true prosperity throughout the kingdom. Our children have to be taught economics and entrepreneurship throughout their childhood, preparing them to assume the role as viable business owners and producers of products, goods and services for the African American community.

Appendix vi

Interview with Howard Starks, lecturer, Wayne State University Department of Africana Studies

Howard Starks, a lecturer at Wayne State University Department of Africana Studies in Detroit makes it a point of duty to invite Black professionals almost on a weekly basis to speak to his class. A devoted teacher of Black studies, Starks had this to say when I gathered his thoughts on Black education.

Bankole Thompson: Where do you see the future of the Black child as far as education is concerned?

Howard Starks: That's a loaded question. If we as African people do not take on the responsibility of teaching our children the much needed information (cultural, political, and capital) that will enable them to maintain and develop a healthy and viable community, we will not have a productive and vigorous future.

What is wrong with the educational system in America today?

The educational system for African Americans is like using gasoline in a diesel engine. Education gives you the information you need to maintain your culture and community; present and future. As long as we keep using information that has no relevance, cultural, geographical, and historical reference—elements of identity; individual and collective, our children will continue to fall victim to those who can relate and operate within the domains of which that information has relevance and meaning.

Why do you invite Black professionals to your class almost on a weekly basis to speak?

Exposure! We need to create cultural learning environments for all students. We need to bring the real world into the classroom so that we can study any and all possibilities for ethnic longevity and prosperity.

Do you think your students gain much from their presentations?

"When the student is ready the teacher will appear." I am obligated and committed to facilitate the learning process for my students. Alas, they are the products of a flawed system; however, we as educators must persistently build on to their porch of knowledge. Their gains can be viewed in the progress of our communities.

Is African-centered education in your opinion given its right of place?

No, Africans in America have been programmed with information that does not represent their way of thinking. They have not been in control of their value system, their political domain, and most certainly, their pedagogical paradigm. As Dubois states, "…this American world—a world which yield him no true self-consciousness, but only lets him see himself through the revelation of the other world." Now is the time to "educate" to educe, to bring forth the sustenance needed to provide our own reparations.

Bibliography

Deborah Bräutigam, Chinese Aid and African Development: Exporting Green Revolution, Palgrave Macmillan, 1998

Walter Rodney, How Europe Underdeveloped Africa, Howard University Press, 1981

Jawanza Kunjufu, An African Centered Response to Ruby Payne's Poverty Theory, African American Images, 2007

Cheikh Anta Diop and Harold J. Salemson, Black Africa: The Economic and Cultural Basis for a Federated State, Lawrence Hill Books, 1987

Charlayne Hunter-Gault, New News Out of Africa: Uncovering Africa's Renaissance, Oxford University Press, 2006

W.E.B. DuBois, The Souls of Black Folk, Penguin Books, 1989

Kevin Gaines, American Africans in Ghana: Black Expatriates and the Civil Rights Era, University of North Carolina Press, 2006

About the author

Bankole Thompson, an award winning journalist, is the **Senior Editor and lead editorial writer** of Michigan Chronicle newspaper Detroit, one of America's leading and oldest Black newspapers founded in 1936. He is also the acting editor of Michigan Front Page, a publication for the young and young at heart.

A widely published journalist, his articles are syndicated by BlackPressUSA.Com, a consortium of 200 Black newspapers reaching 15 million readers weekly. His articles appear in respected media publications across the country such as the Amsterdam News, The Pittsburgh Courier, The Wilmington Journal, The Chicago Defender, The Seattle Medium, Washington Informer, Sacramento Observer, Kansas City Call, South Carolina Black News, Pacific News, etc.

His articles are also documented at the index of International Black Periodicals.

Bankole provides insightful analysis on global socio-political and economic issues, especially as they relate to the Third World. In 2003, he directed the Africa Faith and Justice Network (AFJN), a Washington DC based foreign policy organization's lecture series at the Brookings Institute, Carnegie

Endowment for International Peace and the National Press Club in Washington DC.

The lectures dealt with US foreign policy and human dimensions of post-conflict reconstruction processes in Africa and Third World countries. Some of his works are archived at Ohio State University's Black Studies department, one of the largest Black Studies programs in the US.

Bankole is also a notable public speaker and media critic described by University of Iowa as "a distinguished lecturer."

He has been invited to speak by many reputable institutions and organizations.

In Detroit, Bankole Thompson is a well known political commentator on many radio and television shows including American Black Journal on WTVS (Channel 56-PBS Affiliate), Flashpoint on WDIV (Channel 4- NBC Affiliate), Spotlight on WXYZ (Channel 7 – ABC Affiliate), Talk Of The Town, on MIX 92.3 FM (Clear Channel Radio), and Wake Up Detroit on WHPR TV 33, 88.1FM.

His first book, *Ignoring The Underprivileged*, published in 2006 is a required reading in some colleges.